Bodies of Water

水无处不在　水的样子

[美] 纳迪亚·希金斯　著

[美] 莎拉·英芬特　绘

[美] 艾瑞克·科斯金恩　作曲

池欣阳　译

中国水利水电出版社
www.waterpub.com.cn
·北京·

U0215890

项目策划：徐丽娟
责任编辑：栾 峰 方 斯
特约编辑：李渝汶
联系方式：luanfeng@mwr.gov.cn 010-68545978

书　　名	水无处不在　水的样子 SHUI WUCHUBUZAI SHUI DE YANGZI
作　　者	［美］纳迪亚·希金斯　著　［美］莎拉·英芬特　绘 ［美］艾瑞克·科斯金恩　作曲　池欣阳　译
出版发行	中国水利水电出版社 （北京市海淀区玉渊潭南路1号D座　100038） 网　址：www.waterpub.com.cn E-mail：sales@mwr.gov.cn 电　话：（010）68367658（营销中心）
经　　售	北京科水图书销售中心（零售） 电话：（010）88383994、63202643、68545874 全国各地新华书店和相关出版物销售网点
排　　版	陆　云
印　　刷	北京尚唐印刷包装有限公司
规　　格	285mm×210mm　16开本　6印张（总）　80千字（总）
版　　次	2022年1月第1版　2022年1月第1次印刷
总 定 价	148.00元（全4册）

凡购买我社图书，如有缺页、倒页、脱页的，本社营销中心负责调换
版权所有·侵权必究

图书在版编目（CIP）数据

水无处不在. 水的样子：汉英对照／（美）纳迪亚
·希金斯著；池欣阳译. -- 北京：中国水利水电出版
社，2022.1
书名原文：Water All Around Us
ISBN 978-7-5226-0105-2

Ⅰ．①水… Ⅱ．①纳…②池… Ⅲ．①水—儿童读物
—汉、英 Ⅳ．① P33-49

中国版本图书馆CIP数据核字（2021）第 210673 号

北京市版权局著作权合同登记号：图字 01-2021-5364

Copyright © 2018 Cantata Learning, an imprint of Capstone All rights reserved.
No part of this publication may be reproduced in whole or in part or stored
in a retrieval system, or transmitted in any form or by any means, electronic,
mechanical, photocopying, recording or otherwise without the express written
permission of the publisher. http://capstonepub.com

This Bilingual Simplified Character Chinese and English edition distributed and
published by China Water & Power Press with the permission of Capstone, the
owner of all rights to distribute and publish same.

本简体中文与英文双语版由中国水利水电出版社发行，版权所有。

亲子学习诀窍

为什么和孩子一起阅读、唱歌这么重要？

每天和孩子一起阅读，可以让孩子的学习更有成效。音乐和歌谣，有着变化丰富的韵律，对孩子来说充满乐趣，也对孩子生活认知和语言学习大有助益。音乐可以非常好地把乐感和阅读能力锻炼有机结合，唱歌可以帮助孩子积累词汇和提高语言能力。而且，在阅读的同时欣赏音乐也是增进亲子感情的好方式。

记住：要每天一起阅读、唱歌哦！

绘本使用指导

1. 唱和读的同时找出每页中的同韵单词，再想想有没有其他同韵单词。
2. 记住简单的押韵词，并且唱出来。这可以培养孩子的综合技能以及英语阅读能力。
3. 最后一页的"读书活动指导"可以帮助家长更好地为孩子讲故事。想一想，音符和歌词里的单词有什么联系？
4. 跟孩子一起听歌的时候可以把歌词读给孩子听。
5. 在路上，在家中，随时都可以唱一唱。扫描每本书的二维码可以听到音乐哦。

扫我听音乐

每天陪孩子读书，是给孩子最好的陪伴。

祝你们读得快乐，唱得开心！

Water pools together to form different bodies of water. Rain falls and snow melts to create splashing streams. Streams turn into a rushing river. Rivers flow into lakes, seas, and oceans. Estuaries form where different bodies of water meet.

小水滴汇集在一起，变成了各种样子。
下雨啦，雪化啦，变成小溪哗啦啦。
小溪小溪跑啊跑，汇成大河滚滚流。
大江大河不停歇，流入湖泊和海洋。
大家一起手拉手，来到河口交朋友。

STREAM

溪流

RIVER

河流

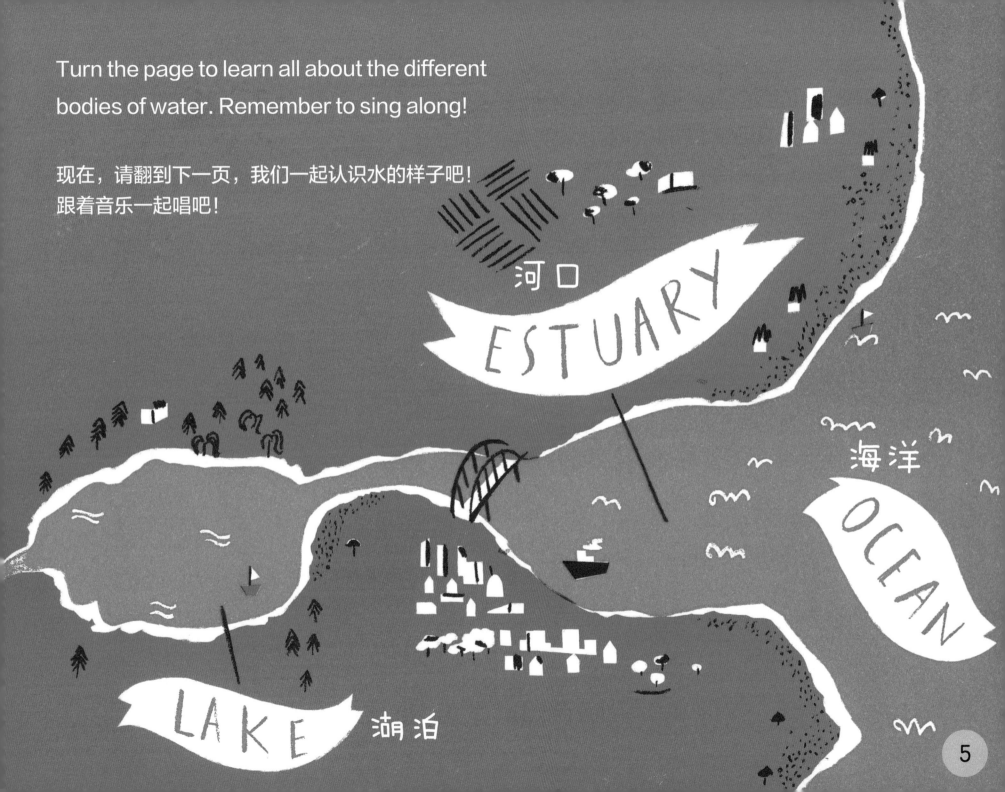

Turn the page to learn all about the different bodies of water. Remember to sing along!

现在，请翻到下一页，我们一起认识水的样子吧！
跟着音乐一起唱吧！

河口
ESTUARY

海洋
OCEAN

LAKE 湖泊

High on a mountain,
a stream, it splashes
down to a river
that ripples and crashes.

高高大山上，小溪哗哗淌，
奔流进大河，水波儿碰撞。

6

The splashing stream,
the rushing river.

哗哗小溪啊，滔滔大河啊。

Bodies of water,

they can flow

from one to another.

Down they go.

水有各种样，永远在流淌，
从你流到我，勇敢向前闯。

That rushing river,

it travels more,

down to a lake

that laps the shore.

滔滔大河流，奔波到永久，
来到湖泊里，拍岸乐悠悠。

The splashing stream,
the rushing river,
the lapping lake.

哗哗小溪啊，
滔滔大河啊，
拍岸湖泊啊。

Bodies of water,
they can flow
from one to another.
Down they go.

水有各种样，永远在流淌，从你流到我，勇敢向前闯。

13

Now from that lake,
the waves, they carry
more water down
to an estuary.

离湖不回头，浪花向前走，
川流不停歇，交汇到河口。

The splashing stream,
the rushing river,
the lapping lake,
the estuary.

哗哗小溪啊，滔滔大河啊，
拍岸湖泊啊，交汇到河口。

Bodies of water,

they can flow

from one to another.

Down they go.

水有各种样，永远在流淌，从你流到我，勇敢向前闯。

17

The estuary,
it swirls into
the open ocean,
so sparkling blue.

相聚河口边，
水花自盘旋，
汇集入大海，
满眼波光蓝。

The splashing stream,

the rushing river,

the lapping lake,

the estuary,

the open ocean.

哗哗小溪啊，
滔滔大河啊，
拍岸湖泊啊，
交汇到河口，
相遇大海中。

Bodies of water,

they can flow

from one to another.

Down they go.

水有各种样，永远在流淌，从你流到我，勇敢向前闯。

SONG LYRICS 歌词
Bodies of Water

High on a mountain,
a stream, it splashes
down to a river
that ripples and crashes.

The splashing stream,
the rushing river.

Bodies of water,
they can flow
from one to another.
Down they go.

That rushing river,
it travels more,
down to a lake
that laps the shore.

The splashing stream,
the rushing river,
the lapping lake.

Bodies of water,
they can flow
from one to another.
Down they go.

Now from that lake,
the waves, they carry
more water down
to an estuary.

The splashing stream,
the rushing river,
the lapping lake,
the estuary.

Bodies of water,
they can flow
from one to another.
Down they go.

The estuary,
it swirls into
the open ocean,
so sparkling blue.

The splashing stream,
the rushing river,
the lapping lake,
the estuary,
the open ocean.

Bodies of water,
they can flow
from one to another.
Down they go.

Bodies of Water

Verse

1. High on a moun-tain, a stream, it splash-es down to a riv-er that rip-ples and crash-es.

The splash-ing stream, the rush-ing riv-er.

Chorus

Bod-ies of wa-ter, they can flow from one to an-oth-er. Down they go.

Verse 2
That rushing river,
it travels more,
down to a lake
that laps the shore.

The splashing stream,
the rushing river,
the lapping lake.

Chorus

Verse 3
Now from that lake,
the waves, they carry
more water down
to an estuary.

The splashing stream,
the rushing river,
the lapping lake,
the estuary.

Chorus

Verse 4
The estuary,
it swirls into
the open ocean,
so sparkling blue.

The splashing stream,
the rushing river,
the lapping lake,
the estuary,
the open ocean.

Chorus

GLOSSARY 词汇表

estuaries—areas where rivers or lakes meet seas or oceans

河口——河流或湖泊与大海大洋交汇的地方

ripples—small waves

水波——小小的水波纹

shore—the land bordering a large body of water

岸——很大一片水体与陆地交汇的边界

streams—flowing bodies of water smaller than a river

小溪——比河流小的流动水体

读书活动指导

1. 想一想在你家附近或者你所在的城市有哪些不同的水体呢？

2. 根据你在这本书里学到的内容，画一种你感兴趣的水体吧。你在上面画了哪些植物或者动物？

3. 向爸爸妈妈、老师、图书管理员要一张你所在地的地图。看到上面的蓝色线条或者蓝色区域了吗？那些就是水域。地图上会标出它们的名字。看看你所在地有哪些水域？